MIGHTY CONSTRUCTION MACHINES

BULLDOZERS

RYAN JAMES

A Crabtree Roots Book

Crabtree Publishing
crabtreebooks.com

School-to-Home Support for Caregivers and Teachers

This book helps children grow by letting them practice reading. Here are a few guiding questions to help the reader with building his or her comprehension skills. Possible answers appear here in red.

Before Reading:

- What do I think this book is about?
    - *I think this book is about how bulldozers work.*
    - *I think this book is about how bulldozers are made.*

- What do I want to learn about this topic?
    - *I want to learn what bulldozers do.*
    - *I want to learn the parts of a bulldozer.*

During Reading:

- I wonder why...
    - *I wonder why some jobs need a bulldozer.*
    - *I wonder why some bulldozers have tracks instead of wheels.*
- What have I learned so far?
    - *I have learned that all bulldozers have a blade.*
    - *I have learned that some bulldozers have four wheels.*

After Reading:

- What details did I learn about this topic?
    - *I have learned that many bulldozers are yellow.*
    - *I have learned that all bulldozers have an engine.*
- Read the book again and look for the vocabulary words.
    - *I see the word **wheels** on page 6 and the word **blade** on page 8. The other vocabulary words are found on page 14.*

This is a **bulldozer**.

Many bulldozers are yellow.

Some bulldozers have four **wheels**.

All bulldozers have a **blade**.

All bulldozers have an **engine**.

How many bulldozers do you see?

CAT

Word List

Sight Words

all
an
are
do
four
have
how
is
many
see
some
this
yellow
you

Words to Know

blade

bulldozer

engine

wheels

29 Words

This is a **bulldozer**.

Many bulldozers are yellow.

Some bulldozers have four **wheels**.

All bulldozers have a **blade**.

All bulldozers have an **engine**.

How many bulldozers do you see?

MIGHTY CONSTRUCTION MACHINES BULLDOZERS

RYAN JAMES

Written by: Ryan James
Designed by: Rhea Wallace
Series Development: James Earley
Proofreader: Melissa Boyce
Educational Consultant: Marie Lemke M.Ed.

Photographs:
Shutterstock: ABCDstock: cover; Maksim Safaniuk: p. 1, 5, 9; Sergei Primakov: p. 3; Dushlik: p. 7; Virrage Images: p/ 11; Alan Budman: p. 13

Crabtree Publishing

crabtreebooks.com 800-387-7650

Printed in Canada/012024/CP20231127

Published in Canada
Crabtree Publishing
616 Welland Ave.
St. Catharines, Ontario
L2M 5V6

Published in the United States
Crabtree Publishing
347 Fifth Ave
Suite 1402-145
New York, NY 10016

Library and Archives Canada Cataloguing in Publication
Available at Library and Archives Canada

Library of Congress Cataloging-in-Publication Data
Available at the Library of Congress

Hardcover: 978-1-0398-3835-2
Paperback: 978-1-0398-3920-5
Ebook (pdf): 978-1-0398-4003-4
Epub: 978-1-0398-4075-1